NUREG-0783

Suppression Pool Temperature Limits for BWR Containments

Generic Technical Activity A-39

U.S. Nuclear Regulatory Commission

Office of Nuclear Reactor Regulation

T. M. Su

Suppression Pool Temperature Limits for BWR Containments

Generic Technical Activity A-39

Manuscript Completed: July 1981
Date Published: November 1981

T. M. Su

Division of Safety Technology
Office of Nuclear Reactor Regulation
U.S. Nuclear Regulatory Commission
Washington, D.C. 20555

ABSTRACT

Boiling water reactor (BWR) plants are equipped with safety/relief valves (SRVs) to protect the reactor from overpressurization. Plant operational transients, such as turbine trips, will actuate the SRV. Once the SRV opens, steam released from the reactor is discharged through SRV lines to the suppression pool in the primary containment. Steam is then condensed in the suppression pool in a stable condition. Extended steam blowdown into the pool, however, will heat the pool to a level where the condensation process may become unstable. This instability of steam condensation may cause severe vibratory loads on containment structures. Current practice in dealing with this phenomenon restricts the allowable operating temperature envelope of the pool in the Technical Specifications so that this instability will not occur. This restriction is referred to as the pool temperature limit. Task Action Plan (TAP) A-39, "Determination of Safety/Relief Valve (SRV) Pool Dynamic Loads and Temperature Limits for BWR Containment," was established to resolve, among other things, the concern about steam condensation behavior for the Mark I, II, and III containments. This report presents the resolution of this issue and includes: (1) the acceptance criteria related to the suppression pool temperature limits; (2) events for which a suppression pool temperature response analysis is required; (3) assumptions used for the analysis; and (4) requirements for the suppression pool temperature monitoring system. This report completes the subtask related to the suppression pool temperature limit in TAP A-39.

CONTENTS

CONTENTS (Continued)

List of Figures

vii

FOREWORD

NUREG-0783 is being issued to provide acceptance criteria for the BWR
suppression pool temperature limit during safety/relief valve discharges to
meet the requirements of General Design Criteria 16 and 29 in Appendix A to
10 CFR Part 50. These criteria are not intended as a substitute for the
regulations, and compliance with them is not required. However, an approach
or method different from these criteria will be acceptable to NRC only if it
provides a basis for determining that these regulatory requirements have been
met.

ACKNOWLEDGMENTS

A-39 REVIEW TEAM

The following individuals participated in Generic Technical Activity A-39, "Determination of Safety/Relief Valve (SRV) Pool Dynamic Loads and Temperature Limits for BWR Containment," and contributed substantially to this report:

T. M. Su, USNRC, Division of Safety Technology (A-39 Task Manager)
A. Sonin, Massachusetts Institute of Technology
C. Economos, Brookhaven National Laboratory
C. C. Lin, Brookhaven National Laboratory
C. Graves, USNRC, Division of Systems Integration

1   INTRODUCTION

Boiling water reactor (BWR) plants are equipped with safety/relief valves (SRVs) to control primary system pressurization. Small pressure variations can be controlled by changing power level and/or load. However, more rapid transients, such as a turbine trips, cannot be handled by such means. For these transients, SRVs mounted on the main steam line are actuated to divert either a portion or all of the steam into the suppression pool. These valves are actuated at individual preset pressure levels or by an external signal (ADS*). The series of SRVs are individually set at pressures over a range so that only the number of valves required to control the pressure transient actuate. Upon SRV actuation, the air column within the SRV discharge line is compressed by the high pressure steam and, in turn, accelerates the water column in the partially submerged line into the suppression pool.

After the water clears from the SRV discharge line, the compressed air also is accelerated into the suppression pool and forms high-pressure air bubbles. These bubbles oscillate between expansion and contraction a number of times before rising to the suppression pool surface.

Following air clearing, essentially pure steam is injected into the pool. Experiments[1] indicate that the steam jet/water interface at the discharge line exit during this phase is relatively stationary when the local pool temperature is low. Thus, the condensation proceeds in a stable manner, and no significant hydrodynamic loads are experienced. Continued steam blowdown into the pool will increase the local pool temperature. The condensation rates at the turbulent steam/water interface are eventually reduced to levels below those needed to readily condense the discharged steam. At this threshold level, the condensation process may become unstable; for example, steam bubbles may be formed and shed from the pipe exit, and the bubbles oscillate and collapse, giving rise to severe pressure oscillations which are imposed on the pool boundaries. Current practice for dealing with this phenomenon in BWR plants is to restrict the allowable operating pool temperature envelope via the Technical Specifications so that the threshold temperature is not reached. This restriction is referred to as "the pool temperature limit."

Task Action Plan A-39, "Determination of Safety/Relief Valve (SRV) Pool Dynamic Loads and Temperature Limits for BWR Containment," was established to resolve, among other things, the concern about steam condensation behavior in the suppression pool of Mark I, II, and III containments. Progress in the resolution of this issue has been reported in NUREG-0661[2] and NUREG-0487.[3] Criteria for the pool temperature limit were established and included in these reports. However, the staff also indicated that the evaluation of this issue would continue in an attempt to improve the criteria, and that further progress would be reported.

This report presents the results of the staff evaluation of the safety issue of suppression pool temperature limits. Acceptance criteria for the pool temperature limits, the events and associated assumptions used to analyze pool temperature response, and the suppression pool temperature monitoring systems are included. The resolution applies to all Mark I, II, and III containments using the SRV quencher devices specified in this report.

---

*
ADS: automatic depressurization system.

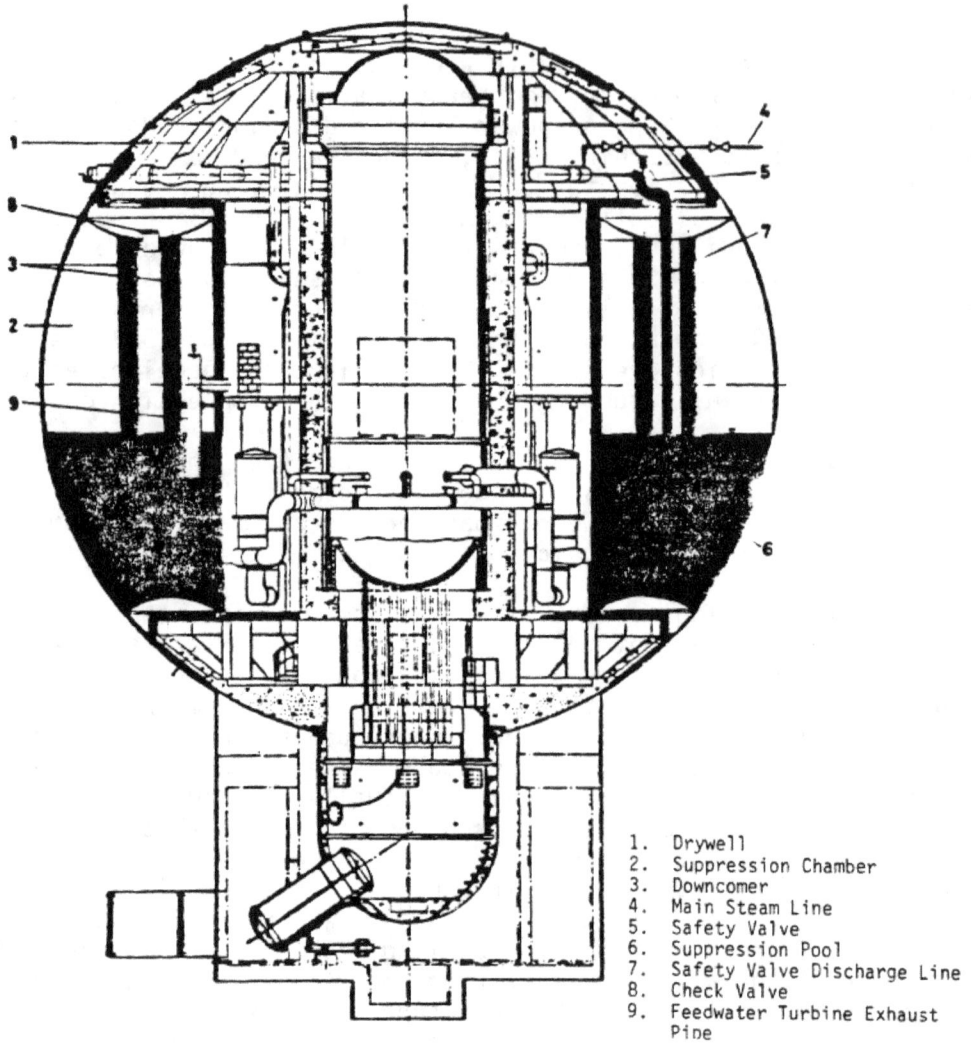

1. Drywell
2. Suppression Chamber
3. Downcomer
4. Main Steam Line
5. Safety Valve
6. Suppression Pool
7. Safety Valve Discharge Line
8. Check Valve
9. Feedwater Turbine Exhaust Pipe

Figure 1   Wurgassen Power Plant, Cross Section
Through the Primary Containment

2   REVIEW OF FIELD EXPERIENCE

2.1 Foreign Plants

2.1.1 Germany[1,4,5]

In April 1972, an incident related to relief valve operation occurred at the
Wurgassen Power Plant (KKW) in Germany. KKW is a BWR plant with a pressure
suppression containment (see Figure 1) and is equipped with eight relief valves.
Each valve is vented into the suppression pool by a vertical pipe approximately
16 inches in diameter that is submerged in 6.5 feet of water.

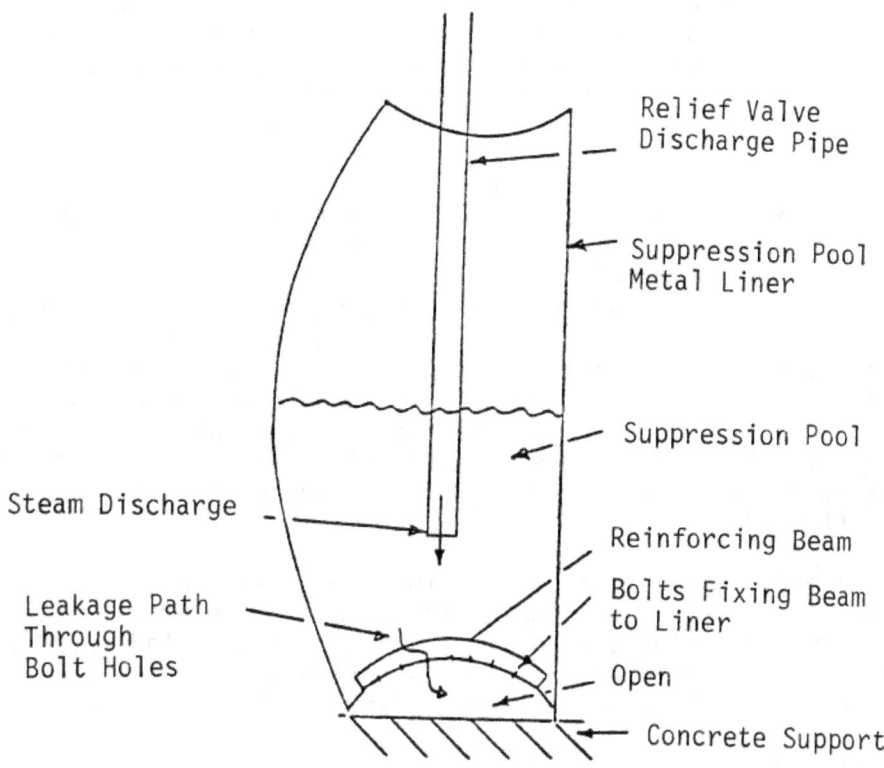

Figure 2   KKW Suppression Pool

During a startup test, relief valves were actuated with the reactor at about 60% of power. One of the relief valves failed to close after being activated for a short period of time. The operator decided to reduce power slowly. During this slow depressurization of the primary system, the suppression pool was gradually heated by the steam released through the stuck-open relief valve. For the first 20 minutes after the valve opened, steam condensation occured smoothly. However, after this initial period when the pool temperature exceeded 160°F, condensation became unstable and the containment structure vibrated severely. The vibration became so severe that the suppression pool metal liner (see Figure 2) separated from the reinforcing beams that had been bolted to the liner. As a result, water leaked through the separation and flowed into the drywell sump.

At the signal initiated by the water leakage, the operator initiated a fast shutdown of the reactor to reduce the pressure of the primary system. The open relief valve closed about 32 minutes after the event was initiated. The reactor then depressurized to ambient conditions, and no further damage to the containment structure occurred. Although the local pool temperatures reached 205°F during this event, the steam condensed completely.

2.1.2 Switzerland

Relief valve tests were performed in July 1972 at a nuclear power plant designed by General Electric (GE) in Switzerland; the plant was at 40% of rated power. Each relief valve had a capacity of about 20% of rated plant power. One relief valve was opened for 5 minutes, at which time an adjoining relief valve was opened. The straight-down discharge pipes were about 47 feet apart. Seven minutes after the opening of the first relief valve, suppression pool vibration was observed. At 8 minutes the test was terminated by closing both valves. Total energy discharged to the pool was 2.6 full reactor power minutes. The pool vibration caused displacement of catwalk sections and the failure of an instrument line, which broke from the suppression pool shell allowing water to flow out.

The estimated local pool temperature between the two relief valve discharges was about 140°F at a discharge mass rate of about 385 lb_m/sec-ft.^{2} Steam condensation was complete during the entire event. This test indicated that pool vibration may be expected when steam condensation occurs with a pool temperature in excess of 140°F and high (400 lb_m/ft^2-sec) SRV discharge rate.

2.2 Domestic Plants

The operating experience of domestic BWR safety/relief valves involving elevated pool temperatures has been summarized in a General Electric report.[6] The following sections, summarized from that report, demonstrate distinct differences between domestic and foreign plant experiences.

2.2.1 Summary of Field Data Survey

A survey of data for SRV discharge events involving elevated pool temperatures was undertaken on operating BWR plants which used ramshead devices (see Figure 3). Of the 11 responses received, five indicated that plants had experienced SRV discharges into suppression pools with temperatures in excess of 100°F with no reported instabilities. These events were as follows:

Plant	Highest Local Pool Temperature, °F
A	165
B	150
C	146
D	129
E	122

2.2.2 Plant A

In Plant A, an SRV stuck open at a reactor pressure of 980 psia, which produced an estimated steam mass flux of 200 lb_m/ft^2-sec to the 10-inch, schedule

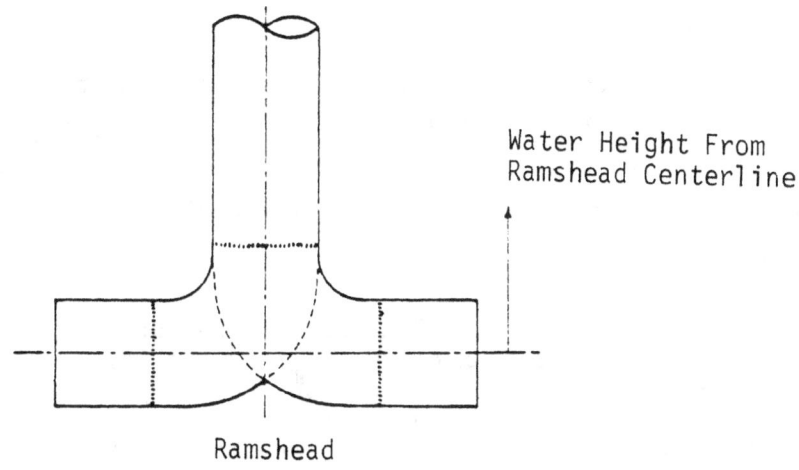

Figure 3 Ramshead Schematic

80 ramshead. Under these conditions, the ramshead discharge mass flux would exceed 40 lb_m/ft^2-sec at reactor pressures in excess of 199 psia (184 psig). The pool temperature, measured deep in the water 90° around the torus from the discharging ramshead, peaked at 165°F before the reactor pressure dropped below 184 psig. The valve eventually reseated at a reactor pressure of 87 psig. A history of the pressures and temperatures is shown in Figure 4. Upon notic- ing that the SRV tailpipe temperature was 400°F, the operators attempted to close the valve. When this proved unsuccessful, torus cooling was initiated with one residual heat removal (RHR) pump and two service water pumps operating. The reactor was manually scrammed when the suppression pool temperature reached 100°F. When the suppression pool temperature reached 165°F, a high-drywell-pressure trip (2 psig) was received, causing the RHR system, standby gas treatment (SBGT) system, and the diesel generators to actuate. A low-water level 2 signal tripped the recirculation pumps and closed the main steam isolation valves (MSIVs). The suppression pool temperature was reduced to < 95°F. When the reactor pressure decreased, emergency core cooling systems (ECCS) were de-activated. The SRV reseated, the RHR torus cooling and torus spray were activated, and the drywell was vented through 2-inch isolation valves to the SBGT system. The reactor then was brought to cold shutdown. No vessel injection by RHR or core spray occurred. Following plant shutdown, visual inspection of the torus exterior revealed no damage.

This blowdown demonstrates that a full-size ramshead device has been operated at a pool temperature higher than 165°F and combined with mass fluxes of more than 40 lb_m/ft^2-sec without observable damage to the torus. The increase in local temperature at the ramshead over that measured 90° around the torus is not known because of the RHR mixing of the pool, but the temperature at the ramshead would not have been less than 165°F. This constitutes the best full-scale evidence to date that the ramshead is capable of operating near 170°F without condensation instability.

2.2.3 Plant B

For Plant B, four separate stuck-open SRV events occurred, producing maximum suppression pool temperatures of 150, 118, 107, and 90°F. During the first

5

event, an SRV failed to reseat during startup tests at 55% of rated reactor power, whereupon the reactor was shut down and depressurized to 350 psig where the SRV reseated. The pool temperature measured 150°F and was 67.5° around the torus from the ramshead. As in Plant A, the local temperature at the ramshead probably was higher than the 150°F at 67.5° around the torus.

2.2.4 Plant C

At Plant C, a SRV stuck open at a reactor pressure of 1005 psia, which caused an estimated steam mass flux of 200 lb_m/ft^2-sec to the ramshead. The reactor was manually scrammed. The suppression pool temperature rose until temperatures located at points 22.5° and 158.5° around the torus from the ramshead reached 140 and 135° F, respectively, as shown in Figure 5. At this time, the reactor pressure dropped below 196 psia (the pressure at which a 10-inch, schedule 80 ramshead would pass a flow rate of 40 lb_m/ft^2-sec). As the pool temperature continued to rise, the reactor pressure dropped to 175 psia, then rose to over 196 psia again before the maximum pool temperature reached 146°F. The SRV reseated at approximately 175 psia reactor pressure 4 hours and 25 minutes after the start of the blowdown. The maximum difference between the two measured temperatures was 10°F, and the difference was 4°F when the temperature peaked.

The torus, inspected visually both externally and internally after this event, showed no evidence of damage.

The local pool temperature at the discharging ramshead was probably slightly higher than the temperatures at the measuring locations but not much more than 4°F higher than the 146 and 142°F peaks measured.

2.2.5 Plants D and E

The responses from Plants D and E contained only statements of the maximum pool temperatures (129 and 133°F) experienced during SRV blowdowns. No data were specifically requested for events involving pool temperatures under 140°F.

2.3 Discussion of Field Experience

A review of the field experience reveals the following distinct differences between the foreign plants and the domestic plants:

2.3.1 SRV Discharge Device

The two foreign plants, which had experienced severe vibration of the containment structures during SRV extended blowdown, used straight-down pipes into the suppression pool as the SRV discharge device. The domestic plants, however, employed ramshead devices. Experimental results[1,6] show that the ramshead device provides a much better steam condensation process than the straight-down pipe.

As shown in Figure 3, the ramshead consists of two elbows welded back-to-back at the end of the SRV discharge line, forming a modified "Y" or "T" junction. At each exit plane, the discharge area is equal to that of the supply pipe. Therefore, the discharge area is increased to twice the supply area.

6

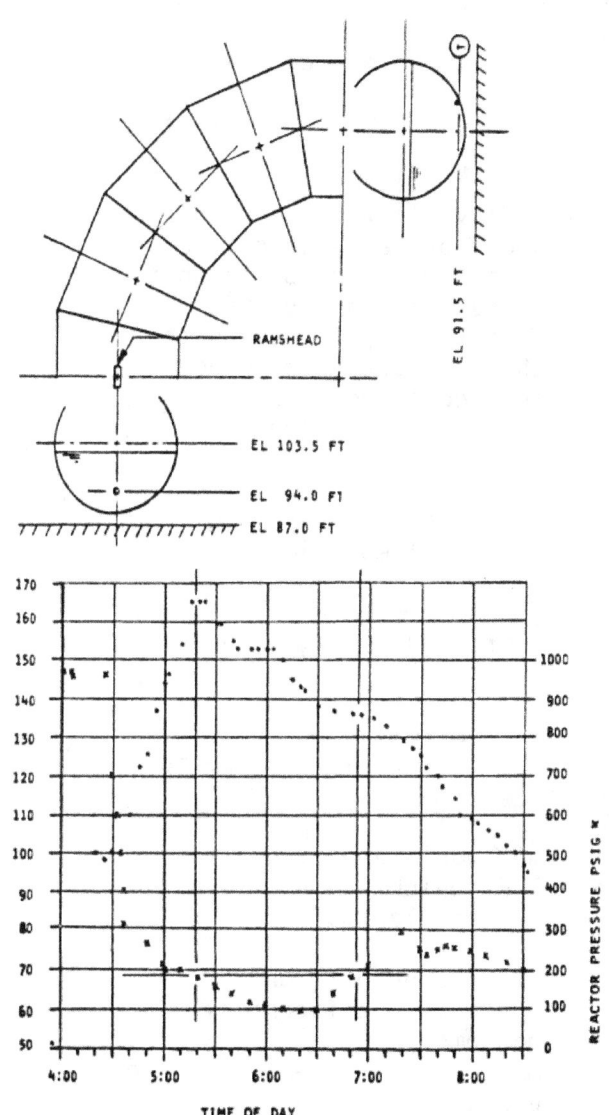

Figure 4   Plant A SRV Blowdown Data

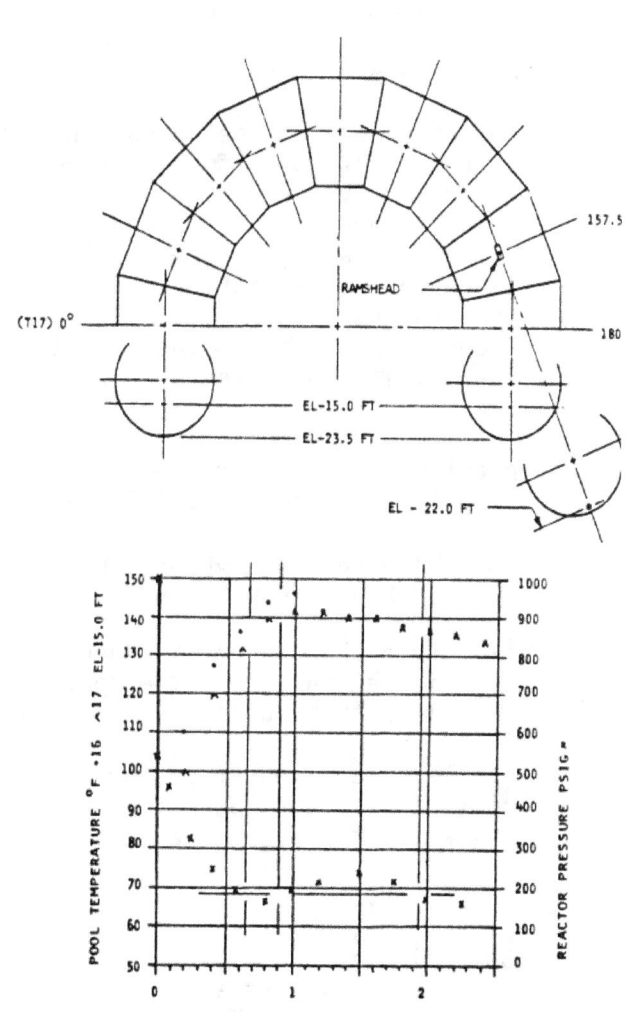

Figure 5   Plant C SRV Blowdown Data

Furthermore, the discharge points are opposite each other along the same axis. This physical arrangement of the ramshead provides better contact between the discharging steam and the pool water. In particular, the horizontal discharge from the ramshead allows rising convection currents and induced secondary flows to circulate cooler water around the steam plumes. This feature of promoting secondary flow results in a higher threshold temperature of condensation instability. Because of this higher threshold, severe vibration has not been observed in the domestic plants which experienced elevated pool temperature during SRV discharge.

2.3.2 Operator Actions

The operators of the Wurgassen power plant intentionally allowed the reactor to remain at power for about 30 minutes while the operators attempted to close the relief valve. During this period, the suppression pool reached the threshold temperature and caused the structural damage. The operators of the domestic plants, on the other hand, had taken prompt action by following the Technical Specifications to scram the reactor and thus the suppression pool temperature rise was minimized. Because the operators complied with the Technical Specifications, the threshold temperature for condensation instability has never been reached at domestic plants.

The Technical Specifications stipulate limiting conditions for operating a BWR plant with a pressure-suppression containment structure. These limiting conditions include the following:

(1) With the reactor at power, the operator shall scram the reactor when the suppression pool temperature exceeds 110°F.

(2) When the suppression pool temperature reaches 120°F following an isolation/scram, the operator shall depressurize the primary system to less than 200 psig at normal cooldown rate.

The Technical Specifications also specify the maximum pool temperature for continuous power operation. In general, this is 95°F, which is considerably below the temperature experienced in the Wurgassen Power Plant.

In summary, the ramshead device with its improved steam condensation capability, combined with the operator actions required by the Technical Specifications, results in improved operation in the domestic plants. However, subscale tests[6] did show that a threshold temperature does exist for the ramshead device. To improve the safety margin, the staff has recommended that all BWR plants with Mark I, II, and III containments use the quencher device (Figures 6, 7, and 8) for SRV discharge. A detailed evaluation of the quencher is presented in the following section.

3   REVIEW OF TEST PROGRAM

3.1 Small-Scale Screening Tests[1]

Results of the Wurgassen incident clearly indicated that the instability of steam condensation during SRV discharge needed further investigation. As a result, a testing program of model discharge devices was initiated at the Mannheim Power Station (GKM) in Germany.

8

The objective of this test program was to develop a discharge device which would greatly reduce or eliminate excessive containment loads during condensation of steam at elevated pool temperatures. Several versions of discharge devices--mainly of perforated pipe arranged in different geometries--were tested. Results of the tests showed a substantial improvement in steam condensation capability, although the results also indicated that a temperature threshold exists for some versions of the perforated pipe. The tests suggest that the hole pattern of the perforated pipe influenced steam condensation and that optimization of the hole pattern was needed to further improve the steam condensation capability.

As a result of this conclusion, tests were performed in the GKM model tank to study the inflow of cooling water for various hole configurations in perforated pipes. By varying the distance between the holes and the diameter of the holes, the effect of the hole pattern was studied. On the basis of this investigation, the hole pattern was optimized for steam condensation.

Because the scaling ratio (1:100 volumetric) is rather large, large-scale tests were needed to substantiate the test results. This phase of quencher device development is discussed in the following section.

3.2 Large-Scale Model Tests[1,7]

This phase of development of the quencher device involved testing various versions of the perforated pipe nozzle under approximate reactor operational conditions. This phase was conducted by large-scale (1:4 volumetric) model tests. These tests were also conducted at the GKM test facilities.

The tests described in References 1 and 7 were subscale in the sense that horizontal and flow-wise cross sections were reduced by a linear factor of 2. The submergence was maintained full scale at about 18.5 feet. The steam flux per unit area was also full scale. The tests were of the "single-cell" type. That is, the quencher was immersed in a water pool of limited area, corresponding to the smallest pool surface which would occur in the prototypical suppression pool. The total perforation area was also reduced by roughly a factor of 4, but the diameter of each hole, as well as the hole spacing, was full scale so as to reproduce the hydrodynamics of the steam/water condensation process near each hole.

The tests of the quencher device with optimized hole pattern[1,7] showed that at high steam mass flow rates (about 95 lb_m/ft^2-sec based on total perforation area), smooth steam condensation was observed with pool temperatures up to about 203°F. This corresponds to a subcooling of 30°F (difference between pool temperature and saturation temperature at the quencher submergence). Tests were also performed at lower mass flow rates (less than or equal to 40 lb_m/ft^2-sec). Again, smooth steam condensation was observed even when the pool temperature reached 214°F. This corresponds to a subcooling of 20°F.

Results of the large scale tests demonstrated that the quencher device performed satisfactorily, with complete condensation of steam for a wide range of conditions anticipated in full-scale plant operations. Because the hole pattern is exactly the same as that being used by the Mark I, II, and III BWR plants, results of the tests can be considered to have direct application for

9

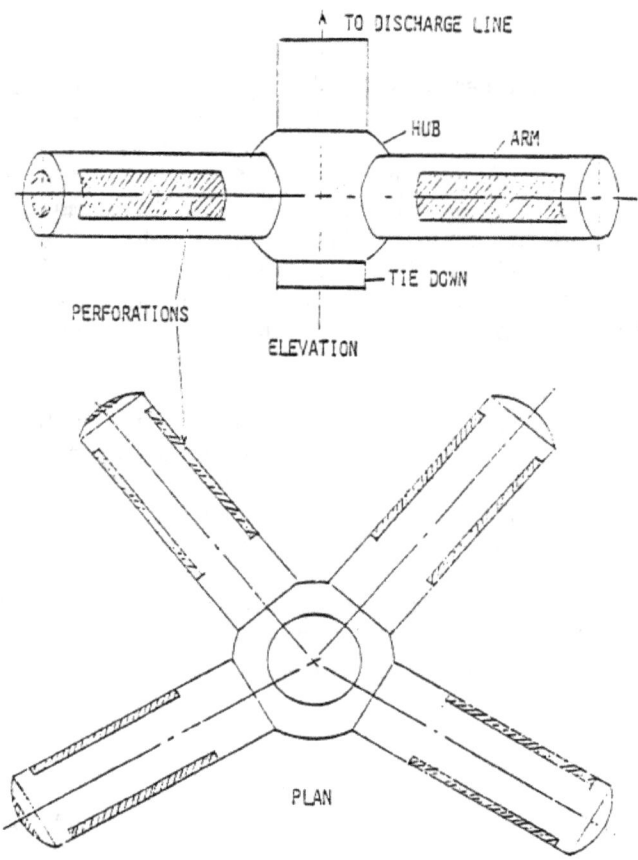

Figure 6   KWU Cross-Quencher

these plants. The data base from these tests forms the base for establishing
the acceptance criteria, which are described in Section 5.1 of this report.

3.3 German Inplant Tests[8,9]

After extensive development tests of the perforated-pipe quencher at GKM (as
described in previous sections), inplant tests were performed in Germany at
the Brunsbuttel Nuclear Power Plant (KKB) in 1974 and the Philippsburg Nuclear
Power Plant (KKP) in 1976. Results of these tests are presented in References 8
and 9.

3.3.1 KKB Inplant Tests[8]

A full-scale version of the KWU X-quencher (see Figure 6) was installed in the
KKB plant. A total of about 100 vent-clearing and condensation tests were
performed with three safety/relief valves. The quenchers were tested over a
wide range of reactor pressure and suppression pool temperatures which
encompassed the plant operating conditions.

Quencher submergence for this plant is 13 feet. The highest tested pool
temperatures ranged from about 150°F at high reactor pressure to about 170°F
at lower reactor pressure. Tests were also conducted at low pool temperature
(about 100°F) throughout the entire range of reactor pressure (75-1100 psi).
Smooth steam condensation was observed.

10

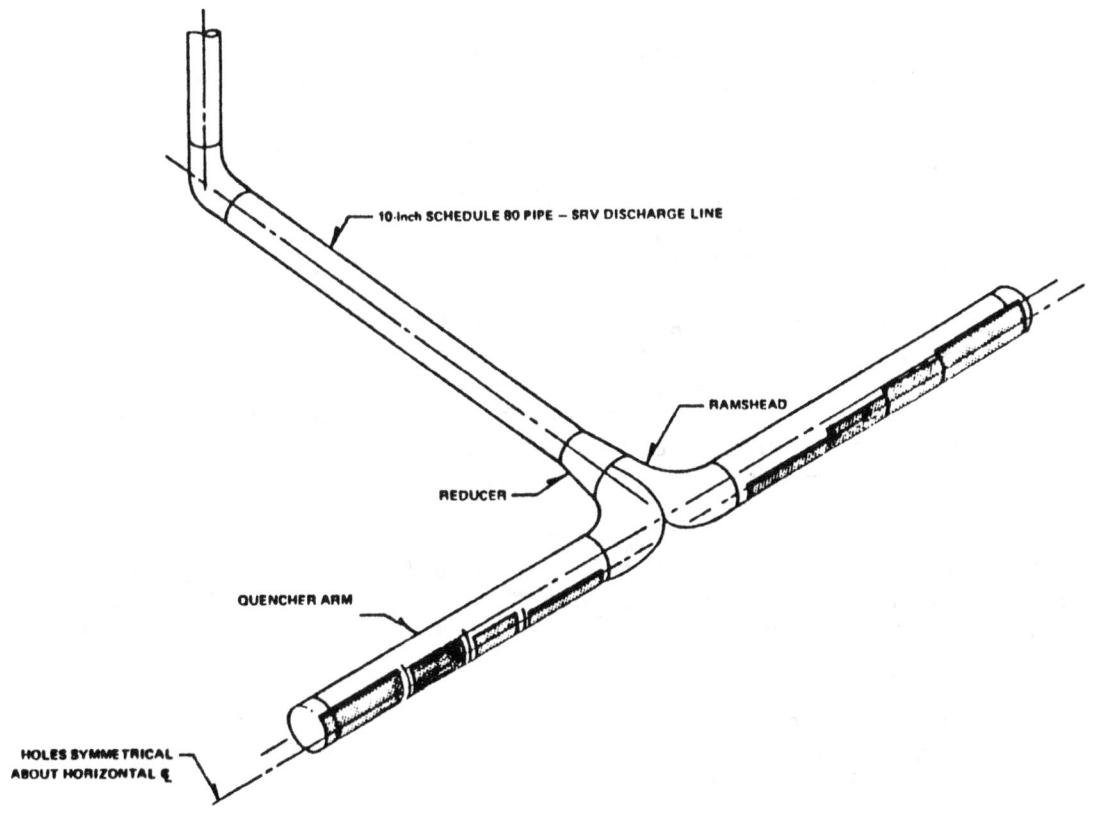

Figure 7 Mark I T-Quencher Shell Discharge Device

The temperature distribution in the suppression pool was nearly uniform, even in longer condensation tests. Thermocouple sensors distributed over the circumference and height of the suppression pool showed that the maximum temperature difference between the highest and lowest reading was about 9°F.

3.3.2 KKP Inplant Tests[9]

The tests conducted in the KKP nuclear plant were similar (except that submergence was 15 feet) but had a somewhat more limited scope than the KKB tests (Section 3.3.1). Specifically, only the low pool temperature range (below 100°F) was examined.

Approximately 70 vent-clearing and condensation tests were performed. The KKP inplant tests demonstrated that the KKB tests are reproducible. KWU personnel concluded, therefore, that further tests with the SRV would not provide new information concerning the loads on the containment and the quencher capability of steam condensation.

3.4 Monticello Inplant Tests[10,11]

In December 1977, the Mark I Owners Group initiated SRV inplant tests at the Monticello Nuclear Generating Station of Northern States Power. The objective of the tests was to evaluate the loads and pool thermal mixing characteristics resulting from SRV discharge through the T-quencher device (see Figure 7).

11

Two extended SRV discharge tests through the T-quencher device were performed at a reactor pressure of about 1000 psia. The first test was performed without the residual heat removal (RHR) system in operation. The second test was similar to the first test except that one loop of the RHR system was in operation throughout the test. The RHR system was in the recirculating mode (no cooling) for this test to determine the possible effect of pool motion induced by RHR operation on the thermal mixing characteristics of the T-quencher. The duration of SRV discharge was about 7 minutes.

Results of the tests show that the maximum bulk to local temperature differences* were 43°F for the test without RHR and 38°F for the test with RHR operation. This represents substantial thermal stratification inside the suppression pool. Because of this significant difference between bulk and local temperature, the Mark I Owners Group decided to perform additional tests with modifications on the T-quencher and RHR discharge.

In February 1978, additional tests of pool thermal mixing were performed at the Monticello plant. The T-quencher device was modified by adding holes to the end cap of one arm of the device. In addition, a 90° elbow with a reducing nozzle was installed at the end of the RHR discharge line. These modifications were made to increase thermal mixing in the pool.

Two tests were performed at conditions similar to those in the previous tests, that is, one with RHR operation and the other without. Evaluation of the test results leads to the following conclusions:

(1) Holes drilled into one end cap of a T-quencher device did induce some pool circulation to enhance thermal mixing. The improvement in the difference between bulk and local temperature, however, is insignificant.

(2) Operation of the RHR system with a modified discharge nozzle resulted in a marked improvement in thermal mixing. The maximum difference in bulk to local temperature was reduced to 15°F, in comparison with 38°F from previous tests without this modification.

(3) The credit for RHR system operation to enhance pool mixing required justification. The test with the RHR system was conducted by having RHR in operation before the SRV was activated. This caused the pool water to move initially at some velocity. Once the SRV activated, the steam released through the SRV was discharged into a swirling rather than a still pool. In actual plant operation, the SRV will initially discharge into a still pool. Only later (about 10 minutes after the pool is heated to the maximum operating temperature) would the RHR system be brought into the pool cooling mode. Additional experimental or analytical justification is needed to confirm the effect of RHR operating time on the bulk-to-local temperature difference.

---

*
See Section 4.2 for the definition.

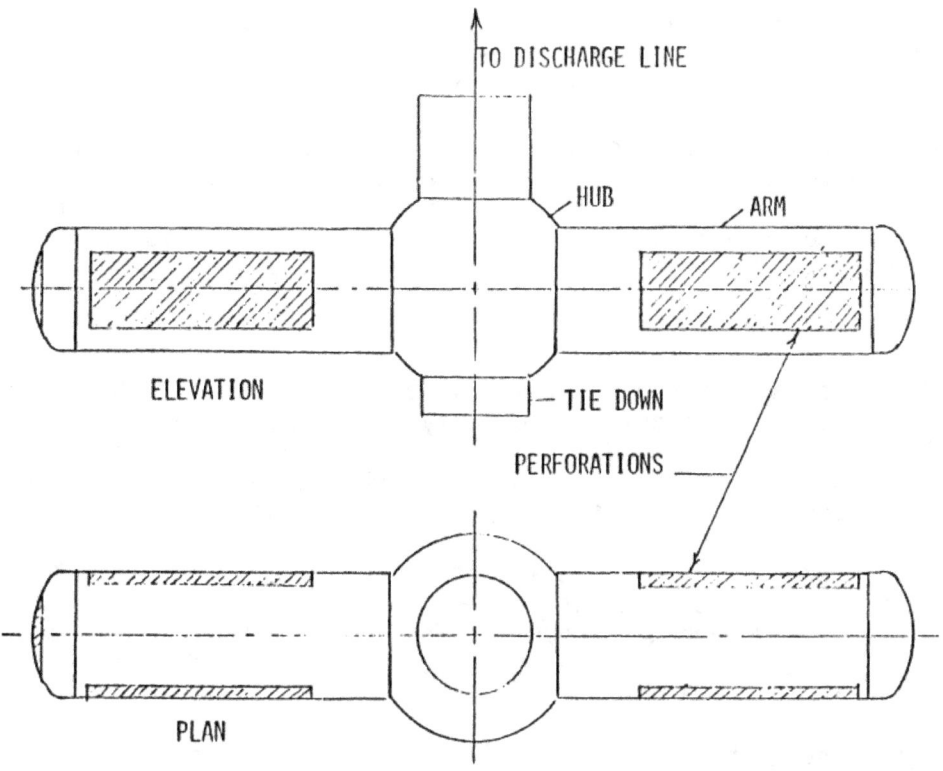

Figure 8   KWU T-Quencher

(4) Because of limitations on pool temperature (105°F) required by the
     Technical Specifications, it is not possible to test the effect of
     RHR operation by following actual operational procedures.*

3.5 Mark II Test Program

3.5.1 KWU T-Quencher Tests[13]

In 1977, Pennsylvania Power & Light Company retained KWU to develop an SRV
discharge device to suit the particular geometry of the Mark II containment.
A two-arm T-quencher (see Figure 8) was subsequently developed and tested at
the Karlstein test facilities in Germany. Detailed evaluation of these tests
will be reported in an NRC staff report which is currently scheduled to be
issued in the third quarter of 1981.

The scope of the steam condensation tests performed is similar to that of tests
conducted in the KKB plant. That is, the reactor pressure and pool temperatures were varied in a way which encompassed the plant operating conditions.

_____
*
 The staff has established acceptable testing procedures that are reported in
 Reference 12.

The results from this program are somewhat more useful in that the conditions encompassed are significantly larger and the test results are reported in more detail. The submergence of the tested device was 21 feet.

The highest tested pool temperature ranged from about 140°F at high reactor pressure (1100 psi) to 196°F at low reactor pressure (30 psi). Blowdown tests were also conducted at low pool temperatures (70 to 90°F) over the entire range of reactor pressures. Smooth steam condensation was observed for the entire test matrix.

3.5.2 Caorso Inplant Tests[14,15]

Inplant safety/relief valve discharge tests were performed at the Caorso Nuclear Station in Italy in late 1978 and early 1979. Detailed evaluation of the tests will be presented in an NRC staff report to be issued late in 1981. This report will address the SRV loads for the Mark III containments which use the Caorso inplant test results as the supporting data base.

The Caorso plant is equipped with an X-shaped quencher similar to the KWU X-quencher (Figure 6), with the exception that there are no holes on the end cap. The quencher was submerged to about 18 feet. There was one extended blowdown test to evaluate the characteristics of pool mixing during SRV discharge.

The test was performed with a uniform initial pool temperature of 60°F. The SRV was actuated and steam was discharged to the suppression pool for about 13 minutes, with the highest pool temperature recorded at about 102°F. The average temperature in the vicinity of the quencher was about 5°F above the calculated bulk temperature at the end of the SRV discharge.

During SRV discharge, the RHR system was not operated in the suppression pool cooling mode. However, the RHR system was activated approximately 5 minutes after the SRV was closed. The data show that pool mixing was accelerated once the RHR was in operation. The temperature measurement indicated that the maximum temperature difference within the pool was about 2°F after 20 minutes of RHR operation. This once again confirms the Monticello inplant test results, which showed a significant improvement of pool mixing by operating the RHR.

3.6 Future Test Program

There are several SRV inplant test programs which are currently underway or are being planned. These include the SRV inplant tests for the Kuosheng Nuclear Power Station, Unit 1, in Taiwan; LaSalle County, Unit 1; and Grand Gulf, Unit 1. SRV extended blowdown tests are included in all test programs.

Both Kuosheng and Grand Gulf are BWR plants with Mark III containment. The SRV tests for the Kuosheng plant are scheduled to be performed in the third quarter of 1981. This plant is to be the world's first Mark III plant in operation. Because all domestic Mark III plants currently have or plan to install the X-shaped quencher, results of the tests should provide valuable information for the assessment of X-quencher performance for Mark III containments.

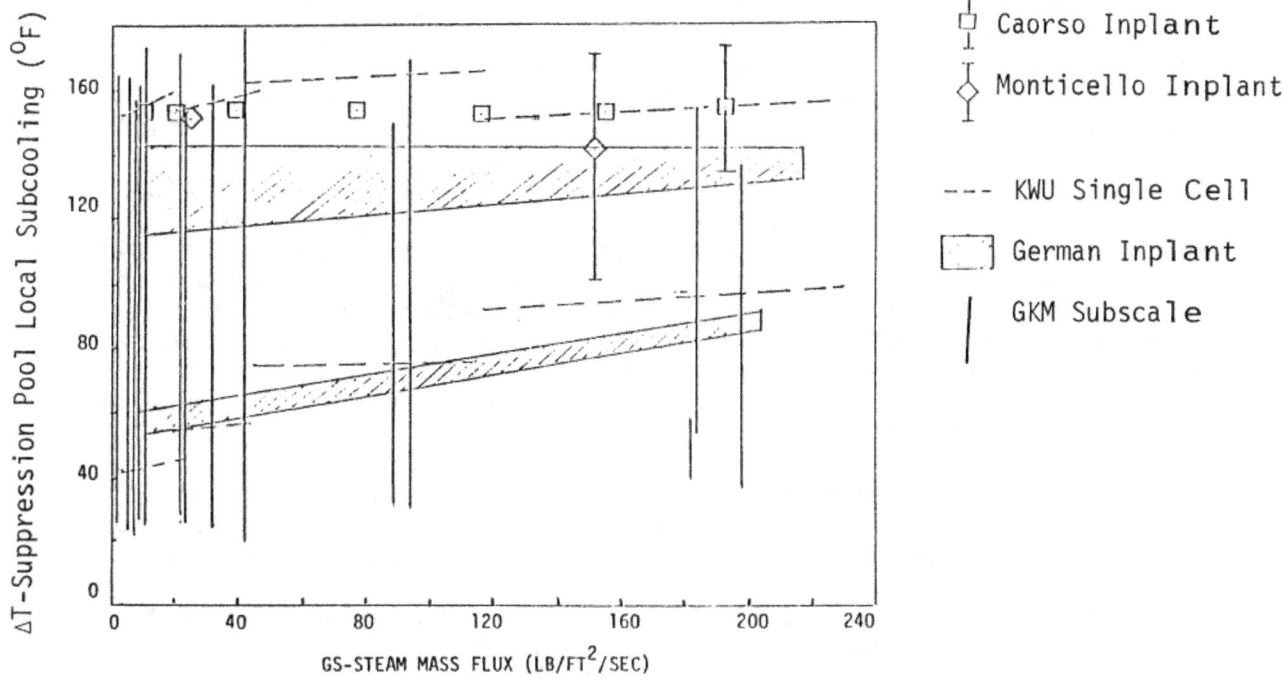

Figure 9   Locus of Points at Which Quencher Steam Condensation
Performance Has Been Observed Experimentally

LaSalle County, Unit 1, is a BWR plant with a Mark II containment. It is
equipped with the KWU-designed T-shaped quencher. Because LaSalle could be
the first Mark II plant in operation in this country, results of the tests
should confirm the T-quencher characteristics as demonstrated at the KWU unit
cell tests. Furthermore, the LaSalle inplant test results will demonstrate
T-quencher performance in a full-size suppression pool rather than in a unit
cell. This includes suppression pool mixing and performance of the temperature
monitoring system during an extended SRV blowdown.

4   EVALUATION OF DATA BASE

The objective of this evaluation is to establish limits on plant operation
during steam blowdown through SRV discharge lines. The limits to be
established can be developed only from the data base discussed in Section 3.

4.1 Operating Limits During Steam Blowdown

Figure 9 summarizes the various tests which were discussed in Section 3.
Because the tests were performed with considerable variation in submergence,

the corresponding values of saturation temperature also vary widely (from about 213 to 237°F). Accordingly, subcooling (ΔT) has been used as the ordinate so that all the tests can be meaningfully grouped. That is, in view of the mechanism associated with condensation instability at high steam flux, ΔT rather than the local pool temperature is considered the more appropriate parameter for characterizing the limits of stable quencher performance.

The information shown in Figure 9 is intended to depict those regions in the ΔT vs steam flux (GS) map where quenchers were actually tested. Within the envelope traced out by the data, it can be anticipated that quencher operation will lead to smooth steam condensation without imposing significant loads on the containment. Quencher operation outside this envelope (that is, at higher values of GS and lower ΔTs) could, in principle, lead to loading conditions significantly more severe than those already observed. However, the actual behavior of SRV discharge operating beyond this envelope is currently unknown because of the lack of experimental evidence.

In the staff's judgment, moderate excursions beyond the envelope are not likely to result in dramatic load increases. Accordingly, quencher operation at the limits of the envelope defined by the existing data base is permissible.

In order to provide a perspective on the adequacy of this operation envelope margin, a typical plant temperature transient has been superimposed on the envelope in Figure 10. In making the transposition, the staff has assumed the use of a Mark II T-quencher, a local-to-bulk temperature difference of 10°F, and a submergence of 13 feet, which corresponds to the shallowest pool among Mark II plants. The correlation between GS and reactor pressure was estimated from GS = 0.28 RP, where RP is reactor pressure in psia. This correlation agrees with the conditions observed during the Caorso tests. Note that uniform application of this linear relation introduces some conservatism at lower values of reactor pressure because the flow rates are overestimated. The comparison shows that a substantial margin exists throughout the transient. Of course, for higher values of local-to-bulk ΔT, this margin could be reduced substantially, particularly at the low end of the GS spectrum. These considerations emphasize the need for applying some judgment in the development of operational limits. For example, the current acceptance criteria for Mark II plants restrict operation to local pool temperatures not greater than 200°F. This limitation is imposed regardless of submergence or steam mass flux. Although this limitation is supported by the data base, it tends to be too restrictive for plants with deep submergence and, in general, for lower steam flux rates. In particular, for values of GS no greater than 42 lb_m/ft^2-sec, a subcooling as low as 20° F can be justified by test data. This would translate into a local pool temperature restriction of 210°F with a 14 ft submergence, thereby providing significant relief in terms of permissible operating transients.

4.2 Local-to-Bulk Pool Temperature Difference

Based on the evaluation presented in Section 4.1, a set of Acceptance Criteria for limiting plant operation has been established to preclude the occurrence of condensation instability dynamic loads. These criteria are presented in terms of limitations on the local pool temperature. In plant transient analyses, however, the bulk pool temperature is used to characterize pool heat-up. Accordingly, the difference between the bulk and local values needs to be specified so that the analysis can demonstrate operation within the prescribed limits.

16

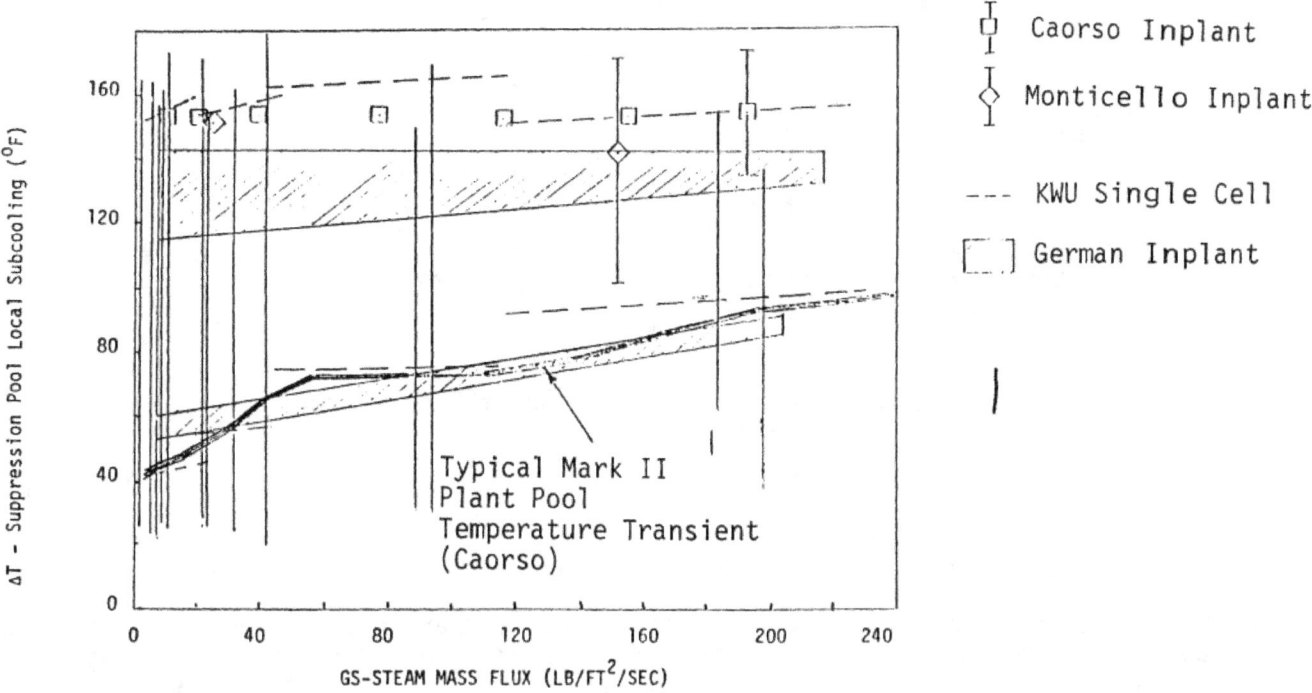

Figure 10   Comparison of a Typical Plant Transient
With the ΔT-GS Map

In the development of the staff criteria, use was made of the data base
obtained from the unit cell tests performed at the GKM facility. In such a
facility, the volume of water associated with a single discharge device is, of
course, only a small fraction of the volume which would exist under prototypical conditions. Because it is a confined pool, differences between local
and bulk conditions are minimal and the temperature recorded by the sensors
can generally be interpreted as local temperatures.* Under inplant conditions, the thermal mixing characteristics will be considerably different.
Thus, some clarification is required regarding the interpretation of local
pool temperature to ensure the correct application of the supporting data base.
This clarification is provided in the next section.

---

*A detailed description of how local temperatures were developed from the GKM
 tests is given in Appendix A.

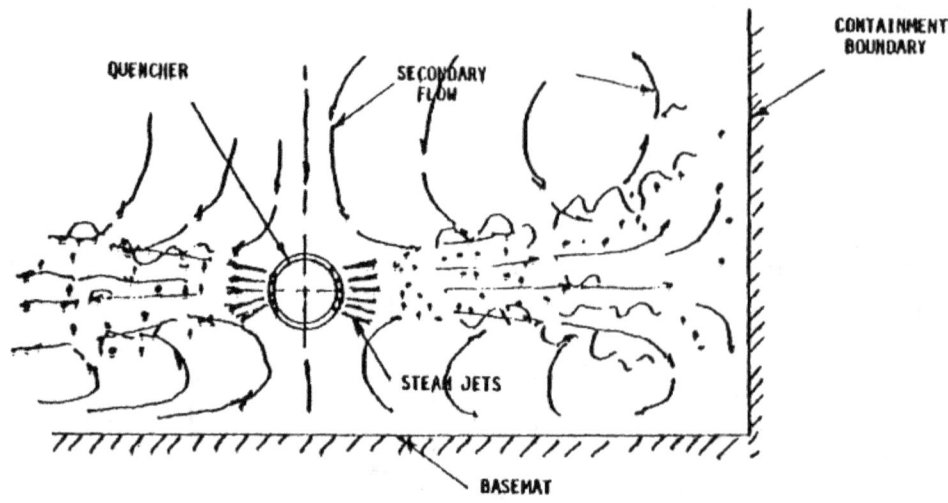

Figure 11   Schematic Representation of Flow Pattern
During Quencher Steam Discharge

4.3   Local Pool Temperature

Local pool temperatures denotes an average water temperature in the vicinity
of the discharge device and represents the relevant temperature which controls
the condensation process occurring at the quencher exit.   In general, this will
differ from both the temperature of the water in contact with the steam and
from the bulk pool temperature.

To define the local pool temperature, a qualitative picture of the flow pattern
during quencher discharge can be evolved by a combination of physical reason-
ing and experimental evidence.   From the data provided by the extensive
temperature array employed in the Monticello SRV tests,[10,11] the general
pattern shown in Figure 11 can be developed.   Assuming that this flow pattern
is a reasonable representation of the actual situation, it is apparent that
the temperature which controls the condensation process (that is, the "local"
temperature) is best characterized by that which would occur at a point
directly above and below the quencher arms (perhaps one or two arms diameters
distant), with the former providing a more conservative measure of this
parameter.

For a variety of practical reasons, temperature sensors cannot always be
located in these optimum locations.   However, the temperature field data
developed from the Monticello tests indicate that, in regions near the boundaries of the containment directly opposite the quencher arms, the temperatures
tend to be somewhat higher than those in the vicinity of the quencher arm.
This is probably the result of the heated plume impinging on these boundaries.
Thus, if local temperatures were defined in terms of such measurements, a
conservatively high value would be deduced.

Some care in applying the local pool temperature defined above is required if
significant pool mixing occurs as a result of RHR operation.   In principle,
this could cause the plume to be swept away from the boundaries thereby
eliminating the conservatism identified above.   Therefore, for plants which

intend to take credit for the effect of RHR operation on local-to-bulk temperature difference, the staff will require that additional temperature sensors be installed on the downstream side (relative to the RHR discharge) of the quencher centerline. The local pool temperature is defined as the average of these values measured at locations directly opposite the quencher arms that are downstream of the RHR discharge.

An averaging of the temperature readings on the boundaries may also be used to define local pool temperature in the absense of the RHR effect. This position is appropriate to ensure that excessive conservatism is not introduced because one particular sensor is directly impinged on by the heated plume.

In applying these averaging techniques, however, certain restrictions will be imposed. These arise from the asymmetry which is exhibited by the temperature field measurements obtained during the Monticello tests. This asymmetry is actually of two types. First there is a clear indication of radial asymmetry, with the highest temperatures consistently recorded in the region between the quencher and the reactor side of the containment. It is the staff's judgment that this is caused by the smaller heat sink capability represented by the water pool which participates in the condensation process on that side (refer to Figure 11). The second type of asymmetry is associated with RHR operation. In this case, the data indicate that the temperatures tend to be higher on the downstream side (relative to the RHR discharge) of the quencher centerline. This is simply a manifestation of the fact that the bulk pool motion has transported water which has been heated by the steam injected through the upstream arm to the vicinity of the downstream arm, where it is heated once again.

In summary, the staff concludes that during quencher operation a bias in the pool temperature field will be created as a result of the annular geometry of the containment. In general, the highest temperature will occur in the region between the quencher and the reactor side of the containment. A temperature bias will also occur during combined RHR and quencher actuation. In this case, the region of the highest temperature will occur on the side of the quencher station downstream of the direction of the RHR discharge.

The acceptance criteria for determining local pool temperatures during inplant testing given in Section 5.0 are based on the considerations outlined above.

5   RESOLUTION OF THE ISSUE

This section presents the technical resolution of the issue regarding suppression pool temperature limits, including the acceptance criteria for the suppression pool temperature limits. The local and bulk pool temperatures are defined. Events which will result in the most severe pool temperature transients also are defined, and the acceptability of the assumptions used to analyze these events is given. Finally, requirements for monitoring the suppression pool temperature are provided.

5.1 Suppression Pool Temperature Limit

The suppression pool temperature limits specified below were established on the basis of the data base discussed in Section 4.1. These limits are applicable only for the quencher device with the exact hole pattern described in Reference 10, 13, or 14. For plants using a discharge device with different hole patterns, applicants and licensees shall provide supporting data to justify the suppression pool temperature limits. To ensure smooth steam condensation without the imposition of significant loads on the containment, the following suppression pool temperature limits shall be used:

(1) For all plant transients involving SRV operations during which the steam flux through the quencher perforations exceeds 94 lb_m/ft^2-sec, the suppression pool local temperature shall not exceed 200°F.

(2) For all plant transients involving SRV operations during which the steam flux through the quencher perforations is less than 42 lb_m/ft^2-sec, the suppression pool local temperature shall be at least 20°F subcooled. This is equivalent to a local temperature of 210°F with quencher submergence of 14 feet.

(3) For plant transients involving SRV operations during which the steam flux through the quencher perforations exceeds 42 lb_m/ft^2-sec but is less than 94 lb_m/ft^2-sec, the suppression pool local temperature can be established by linearly interpolating the local temperatures established under items (1) and (2) above.

5.2 Local-to-Bulk Pool Temperature Difference

(1) The local-to-bulk* pool temperature difference shall consider the plant-specific quencher-discharge geometry and RHR suction and discharge geometry.

(2) Determination of the plant-specific local-to-bulk pool temperature difference shall be supported by existing pool temperature data or by additional inplant tests.

(3) Where inplant tests are used to establish the local-to-bulk pool temperature difference, the test procedures and instrumentation shall be in accordance with the guidelines specified in NUREG-0763.[12]

5.3 Local Pool Temperature

The local pool temperature is defined as the fluid temperature in the vicinity of the quencher device during steam discharge. For practical purposes, the average water temperature observed within the region subtended by the quencher arms on the reactor side of the containment and at the same elevation as the quencher device can be considered the local temperature. For plants for which credit is to be taken for the effectiveness of the RHR to mix the pool, local temperature can be defined by using the average temperatures measured by only the sensors downstream of the quencher (relative to the RHR flow). However, in this case also, temperatures directly above the downstream quencher arm shall be included in the averaging.

---

* See Section 5.4 for definition.

5.4 Bulk Pool Temperature

The bulk pool temperature is the temperature calculated by plant transient analyses assuming that the suppression pool acts as a uniform heat sink. Bulk temperature is calculated on the basis of mass and energy released from the primary system through the SRVs after plant transients.

5.5 Local Subcooling

Local subcooling is defined as the difference between the local pool temperature and the saturation temperature, corresponding to the hydrostatic pressure at the quencher elevation and atmospheric pressure. Pressurization of the containment atmosphere above normal atmospheric pressure shall not be considered for determining local subcooling.

5.6 Events for the Analysis of Pool Temperature Transients

The operational temperature limit established for the suppression pool is described in Section 5.1. To meet this limit, applicants and licensees are required to provide an analysis for suppression pool temperature response to various SRV events. These events can be analyzed on the basis of mass and energy balance on the suppression pool during SRV blowdown. Results of the suppression pool temperature transient will demonstrate the history of the bulk temperature of the suppression pool for all the events analyzed. Whether the plant meets the limit can be clearly demonstrated by the difference between local and bulk temperatures as described in Section 5.2. The events required to be analyzed in accordance with assumptions prescribed in Section 5.7 are as follows:

5.6.1 Stuck-Open SRV (SORV) During Power Operation

This event postulates that an SRV is inadvertently actuated while the plant is operating at power, as defined in Section 5.7.1. After to the activation, the SRV fails to reseat and remains open. As a result of this malfunction, steam from the primary system is discharged through the SRV and released to the suppression pool. The following two cases shall be analyzed for this event:

(1) Loss of one RHR heat exchanger

(2) Initiation of the main steam isolation valve (MSIV) closure signal at the time of scram

5.6.2 SRV Discharge Following Isolation/Scram

This event postulates that a sudden closure of the MSIVs and subsequent scram occur in response to plant operational transients. SRV discharge is required to depressurize the reactor. The rate of reactor depressurization shall follow the assumption specified in Section 5.7.2.2. Note that this case is equivalent to the scenario that postulates a stuck-open SRV following the isolation and scram with loss of one RHR heat exchanger. This results since the peak pool temperature occurs late in the transient, typically 2 to 3 hours after reactor scram, and an equivalent amount of energy is transferred to the suppression pool for both cases.

5.6.3 SRV Discharge Following a Small-Break Loss-of-Coolant Accident (SBLOCA)

This event postulates that a small-break accident occurs in the primary system. SRV discharge is required to depressurize the reactor coolant system and then remains open. Loss of one RHR heat exchanger shall be assumed.

5.7 Assumptions Used in the Analysis

5.7.1 General Assumptions

The following general assumptions should be used for all cases described in Section 5.6:

(1) The power level, decay heat, service water temperature, RHR heat exchanger capability, and suppression pool initial temperature are consistent with those used for the analysis of containment pressure and temperature response to a loss-of-coolant accident. These values are specified in Chapter 6.2 of PSARs or FSARs.

(2) The initial water level of the suppression pool is at the minimum level indicated in the Technical Specificiation.

(3) Main Steam Isolation Valve (MSIV) closure is complete 3.5 seconds after the isolation signal for transients where isolation occurs.

(4) The water volume within the reactor vessel pedestal (Mark II) or within the weir wall (Mark III) is not included in the calculation of pool temperature response.

(5) Feedwater pumps supply feedwater to the reactor until the feedpumps trip on an automatic signal. The applicant or licensee is required to provide information regarding the history of feedwater mass and energy addition to the reactor.

(6) Offsite power is not available for isolation/scram and small break loss-of-coolant accidents, except that offsite power is available for feedwater pumps.

(7) The high-pressure core injection (HPCI) or high-pressure core spray (HPCS) systems are terminated at the specified high pool temperature. This temperature may be a plant-unique specification and should be provided by the applicant or licensee.

(8) The applicant or licensee provides information to demonstrate that no single failure, either in the system design or power source, will result in the loss of one RHR heat exchanger and the RHR shutdown cooling mode.

(9) Calculation of mass and energy release to the suppression pool through SRV shall follow the methodology described in Reference 17.

(10) There are no heat losses to the containment atmosphere and structures.

(11) The RHR operates in the suppression pool cooling mode 10 minutes after TS1* is exceeded. However, the operation of pool cooling may be interrupted by

---

* TS1 is the Technical Specification maximum pool temperature for continued power operation.

other requirements. For instance, high drywell pressure may automatically terminate the RHR pool cooling mode. Provide instructions in the operating procedures to reinstate the pool cooling mode. Identify the duration of this interrupted pool cooling mode and include it in the analysis for suppression pool temperature response.

5.7.2 Assumptions for Specific Events

This section discusses the specific assumptions used for the events described in Section 5.6. It also provides restrictions and guidance for justification for the assumptions.

5.7.2.1 SORV at Power

(1) Case 1.a

Assume that manual scram can be accomplished when the suppression pool temperature reaches 110°F as indicated in the Technical Specifications. To justify this assumption, the applicant or licensee shall meet the following requirements:

(a) Install alarms/displays in the control room to give the operator immediate and unambiguous indications of a stuck-open SRV.

(b) Provide alarms/displays to alert the operator about the suppression pool temperature. Set the alarm at TS1 and TS3.*

(c) Provide clear instructions in operating procedures to prohibit the operator from prolonging the initiation of manual scram. For example, the operational procedures should specify the maximum number of attempts the operator will be allowed to use to reclose a stuck-open SRV.

If the applicant or licensee does not meet all of these requirements, manual scram shall be assumed to be accomplished 10 minutes after the pool temperature reaches 110°F.

Assume the main condenser is available as an alternative heat sink. The use of the main condenser as a heat sink requires that the bypass system be available, that the circulating water system function, and that the main steam isolation valves remain open. The applicant or licensee, therefore, is required to provide information that demonstrates the availability of these systems. Furthermore, the applicant or licensee is required to justify that containment-structure isolation is not required for the event.

(2) Case 1.b

Assumptions used for Case 1.a above are applicable for this case, with the following exceptions:

Assume the main condenser is not available because of spurious MSIV closure.

---

* TS3 is the Technical Specification pool temperature limit for reactor scram.

Assume two RHR heat exchangers are available.

5.7.2.2 SRV Discharge Following Isolation/Scram

(1) Assume the loss of one RHR heat exchanger.

(2) Assume the RHR is operating in the suppression pool cooling mode 10 minutes after isolation or scram is initiated. Meet the requirements specified in Section 5.7.1(11) for operational procedures and the duration for interruption of RHR operation.

(3) Following reactor isolation or scram, assume that manual depressurization can be initiated at a pool temperature of 120° F, which is the plant Technical Specification for reactor depressurization.

(4) Assume the rate of manual depressurization can be controlled at the normal rate of 100° F per hour. If higher cooldown rates are to be used, justification shall be provided.

5.7.2.3 SRV Discharge Following a Small-Break Accident

(1) Apply the assumptions described in Section 5.7.2.2.

(2) Perform a sensitivity analysis to determine the effect of the loss-of-shutdown-cooling mode on the suppression pool heatup rate.

(3) Assume the reactor is scrammed on high drywell pressure.

(4) Assume the MSIV closure signal is activated at the onset of the accident.

5.8 Suppression Pool Temperature Monitoring System

The suppression pool temperature monitoring system is required to ensure that the suppression pool is within the allowable limits set forth in the plant Technical Specifications. The system shall meet the general design requirements listed below. It should be noted that specific criteria provided in Reference 3 shall be used for the Mark II plants.

(1) Each applicant or licensee shall demonstrate adequacy of the number and distribution of pool temperature sensors to provide a reasonable measure of the bulk temperature. Alternatively, redundant pool temperature monitors may be located at each quencher, either on the quencher support or on the suppression pool wall, to provide a measure of local pool temperature for each quencher device. In such cases, the limits for pool temperature shall be derived from the calculated bulk pool temperature and the bulk to local pool temperature difference transient.

(2) Sensors shall be installed sufficiently below the minimum water level, as set forth in the plant Technical Specifications, to ensure that the sensor properly monitors pool temperature.

(3) Pool temperature shall be indicated and recorded in the control room. Where the suppression pool temperature limits are based on bulk pool temperature, operating procedures or analyzing equipment shall be used to

minimize the actions required by the operator to determine the bulk pool temperature. Operating procedures and alarm set points shall consider the relative accuracy of the measurement system.

(4) Instrument set points for alarms shall be established so that the plant will operate within the suppression pool temperature limits discussed above.

(5) All sensors shall be designed to seismic Category I, Quality Group B standards, and shall be capable of being energized from onsite emergency power supplies.

6   IMPLEMENTATION

The staff's recommendations for implementing the technical resolution described in Section 5 are summarized in the paragraphs below.

6.1   Implementation of the Resolution for Mark I, II, and III Plants

The staff recommends that a generic letter and a copy of this report be sent to licensees and applicants with Mark I, II, and III containments requesting the information described in Section 5. For a response to be acceptable to the staff, it must conform to the requirements specified in Section 5.

It should be noted that acceptance criteria for the suppression pool temperature limits have been issued and presented in NUREG-0661[3] (Mark I) and NUREG-0487[4] (Mark II), and licensees or applicants may follow those criteria. However, those criteria are more restrictive than the criteria specified in this report. Specifically, NUREG-0661 and NUREG-0487 specify a fixed pool temperature limit (200°F) for all operating conditions. The pool temperature limits presented in this report have taken plant-specific geometries (quencher submergence) and plant operating conditions (steam flux through the quencher device) into consideration. As a result, the pool temperature limits can be relaxed for certain plants.

The resolution of the pool temperature limit issue has been developed to be as generic as possible. Plant-specific review, however, is still required.

Those areas which require plant-specific review are identified in Sections 5.6 and 5.7.

6.2   Recommended Changes to Standard Review Plan

The following changes to SRP Section 6.2.1.1c are recommended:

I. Areas of Review:

11. Suppression pool temperature limit during safety/relief operation, including events to analyze suppression pool temperature response, assumptions used for the analysis, and the suppression pool temperature monitoring system.

II. Acceptance Criteria

9. For Mark I, II, and III plants, the suppression pool temperature should not exceed 200°F or the acceptance criteria specified in Section 5.1 of NUREG-0783.

III. Review Procedure

9. The Containment System Branch (CSB) will review the information related to the suppression pool temperature limit during safety/ relief valve operation involving either normal plant transients and LOCA of small line breaks (those events specified in Section 5.6 of NUREG-0783). The CSB will evaluate the difference between bulk and local temperatures defined in Section 5.3 of NUREG-0783 and the suppression pool temperature monitoring system described in Section 5.8 of NUREG-0783. With respect to the assumptions used to analyze the events, the CSB will coordinate its review with the Reactor Systems and Auxiliary Systems Branches to review the assumptions related to the emergency core cooling and residual heat removal systems, and main-condenser-related systems, such as the main-condenser cooling system. Acceptability of the suppression pool temperature limit shall be based on conformance with the resolution of the issue specified in Section 5 of NUREG-0783.

6.3 Recommended Development of Regulatory Guide

A Regulatory Guide should be developed to provide guidance for applicants and licensees to deal with the issue related to safety/relief valve dynamic load and pool temperature limits. With respect to suppression pool temperature limits, Sections 5.1 through Section 5.8 of this report should be included in the Regulatory Guide.

7 REFERENCES

References cited in this report are available as follows:

Those items marked with one asterisk (*) are available for inspection in the NRC Public Document Room, 1717 H Street, N.W., Washington, D.C. 20555; they may be copied for a fee.

Material marked with two asterisks (**) is not publicly available because it contains proprietary information; however, a nonproprietary version is available in the NRC Public Document Room for inspection and may be copied for a fee.

Those reference items marked with three asterisks (***) are available for purchase from the NRC/GPO Sales Program, U.S. Nuclear Regulatory Commission, Washington, DC 20555, and/or the National Technical Information Service, Springfield, Virginia 22161.

All other material referenced is in the open literature and is available through public technical libraries.

1. Fukashima, T. Y., et al., "Test Results Employed by General Electric for Boiling Water Reactor Containment and Vertical Vent Loads," GE Report NEDE-21078P, October 1975.**

2. U.S. Nuclear Regulatory Commission, "Safety Evaluation Report: Mark I Containment Long-Term Program, Resolution of Generic Technical Activity A-7," USNRC Report NUREG-0661, July 1980.***

3. U.S. Nuclear Regulatory Commission, "Mark II Containment Lead Plant Program Load Evaluation and Acceptance Criteria, "USNRC Report NUREG-0487, October 1978, and Supplement No. 1, September 1980.***

4. General Electric, "Suppression Pool High Temperature Steam Quenching Vibrations," December 1975.*

5. Voigt, O., et al., "Consequences drawn from a Stuck Relief Valve Incident at the Wurgassen Power Plant," presented at the 2nd International Conference on Structural Mechanics in Reactor Technology, September 10-14, 1973, in Berlin.

6. General Electric Company Memorandum Report, "170° F Pool Temperature Limit for SRV Ramshead Condensation Stability," September 1977.**

7. KWU, "Condensation and Vent Clearing Tests in GKM with Perforated Pipes," KWU Technical Report KWU/E3-2594, May 1973.**

8. KWU, "Results of the Non-Nuclear Hot Tests with the Relief System in The Brunsbuttel Nuclear Power Plant," KWU Technical Report KWU/R 113-3267, December 1974.**

9. KWU, "Results of the Non-Nuclear Hot Test with the Relief System in the Philippsburg Nuclear Power Plant," KWU Working Report R142-38/77, March 1977.**

10. Asai, R. A., and others, "Mark I Containment Program Final Report, Monticello T-Quencher Test," GE NEDE-21864-P, July 1978.**

11. General Electric Company, "Mark I Containment Program - Monticello T-Quencher Thermal Mixing Test Final Report," GE NEDE-24542-P, April 1979.**

12. U.S. Nuclear Regulatory Commission, "Guidelines for Confirmatory Inplant Tests of Safety-Relief Valve Discharges for BWR Plants," USNRC Report NUREG-0763, May, 1981.***

13. Pennsylvania Power and Light Company, "Design Assessment Report, Susquahanna Steam Electric Station, Unit 1 and 2, Section 8.0," Revision 1. March, 1979.**

14. General Electric Company, "Mark II Containment Supporting Program - Caorso Safety Relief Valve Discharge Tests - Phase II Test Report," GE NEDE-24757-P, May 1980.**

15. General Electric Company, "Mark II Containment Program - Caorso Extended Discharge Test Report," GE NEDE-24798-P, July 1980.**

16. KWU, "Investigation of Condensation with the Perforated-Pipe Quencher with Small Water Coverage of the Quencher Arms," KWU Technical Report KWU/E3-2840, December 1973.**

17. Letter report, R. H. Bucholz to Karl Kniel dated March 12, 1981, "Mark II Containment Program Method for Calculating Mass and Energy Release for Suppression Pool Temperature Response to Safety Relief Valve Discharges."*

Appendix A - Description of GKM Large-Scale Perforated-
Arm Quencher Test Program

The steam-condensing performance of perforated-pipe quenchers to be used for
mitigation of SRV loads was evaluated during a series of tests performed at
the GKM facility by KWU.[1] The primary objective was development of an optimum
hole pattern for the perforations which would ensure stable steam condensation
at elevated suppression pool temperatures.

The test facility used during this investigation simulated a small sector of a
BWR wetwell (see Figure A-1). The facility consisted of cylindrical tank
9 feet in diameter and 60 feet high, partially filled with water. A single
SRV discharge line 8 inches in diameter was concentrically installed in the
tank. The perforated arms forming the quencher were attached to the end of
this discharge line approximately 4 feet above the pool bottom. The arms were
arranged in a generally cruciform manner but were inclined at 45° below the
horizontal. The depth of the pool was varied from a minimum of 15 feet to a
maximum of 23 feet.

The test procedure consisted of continuous blowdown of steam through the
discharge line into the suppression pool at a fixed flow rate until the pool
water temperature approached saturation conditions. Twenty-six such blowdowns
were performed with steam flux rates ranging from 2 to 128 lb_m/ft^2-sec, based
on total perforation area. The 26 blowdowns were performed with seven different versions of perforations. Typical parameters varied were the hole
diameters and the vertical and horizontal spacing between the rows and columns
of holes. An additional variation included the use of perforations on the
discharge line itself. In some cases these perforations were located as far
as 9 feet above the quencher arms where the bulk of the perforations were
located. During the blowdowns, the pressure and fluid temperature at the pool
boundaries were monitored continuously by means of sensors installed for that
purpose. In particular, the onset of instabilities in the condensation process was determined by noting if the pressure amplitudes exhibited a tendency
to increase in a more or less discontinuous manner as the pool water approached
saturation conditions. These observations formed the criteria by which the
suitability of any particular quencher design was established.

Of the seven versions which were tested, only three performed acceptably in
the sense indicated above. All of these had an identical arrangement of
perforations on the quencher arms themselves. They differed only with respect
to the perforations on the discharge line. Twelve blowdowns were performed
with these optimized versions. The submergence during these blowdowns (that
is, the vertical distance between the center perforations on the quencher arms
and the pool surface) was fixed at about 18-½ feet. This corresponds to a
local saturation temperature of 234°F.

Pressure and temperature data are reported for eight of these blowdowns. The
pressure data, presented graphically, show the variation of peak pressure
amplitude with pool temperature. The former are characterized by the peak
pressures recorded by two sensors located at the tank bottom. One of these
was coincident with the tank axis (sensor P5) and was approximately 4-½ feet
below the center of the perforations.

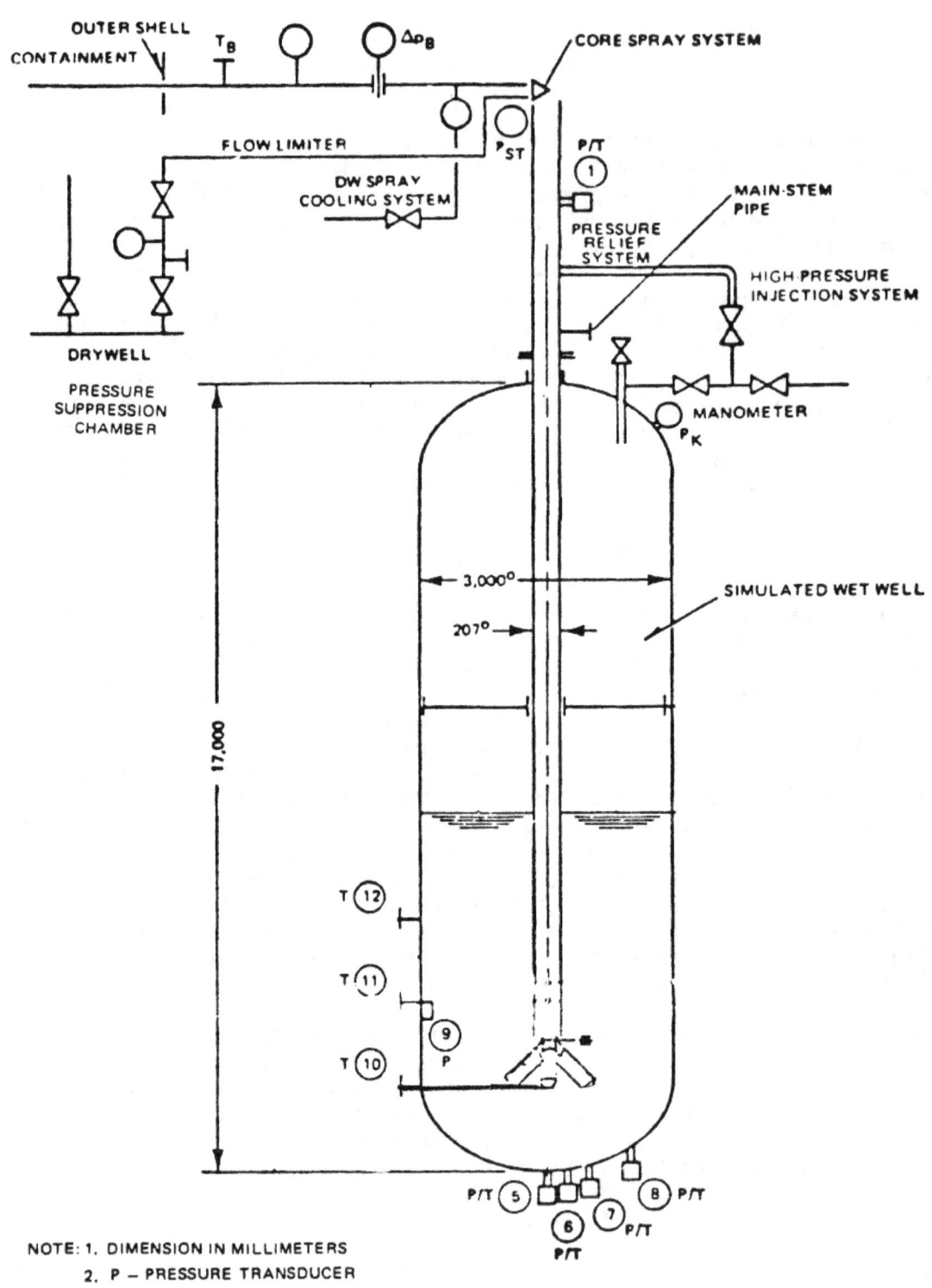

Figure A-1 Large-Scale Test Facility

The other sensor was located at a radial distance 1-½ feet from the tank axis; it was in the symmetry plane passing between adjacent arms (Sensor P71). Although other pressure sensors were available, only the pressures recorded by these sensors were reported because they were the only ones with an external sensing diaphragm and, therefore, free of fluid structure interaction effects.

The pool temperature was characterized in the graphical results by the output from sensor T11, which was on the tank wall 6-½ feet above the tank bottom. Sensor T11 coincided with a plane passing through the tank axis forming an angle of 15° with a plane passing through the axis of the quencher arms.

The temperatures recorded by sensor T11 at the beginning and end of each blowdown are also reported in tabular form by test number. For five blow-downs, additional temperature data are also presented graphically.[1,2] These plots show both the time and spatial distribution of the temperature field throughout the submerged boundaries of the test tank.

A total of 14 sensors made up the array which defined this temperature field. Of these, seven were on the tank bottom, and the remainder on the tank walls in two distinct planes. One of these planes was coincident with a symmetry plane passing between the quencher arms. Three sensors (T52, 53, 55) were located here, 5.7, 7.4, and 9.8 feet above the pool bottom. The remaining four sensors included sensor T11 whose location is described above. The other three sensors (T10, T12, and T13) were located 3.3, 9.8, and 13.1 feet above the pool bottom in the same plane as T11.

In Section 5.1 of NUREG-0783, operational limits for the T-quencher were established in terms of a local subcooling (ΔT), defined as the difference between saturation temperature at the quencher location and local pool temperature. Local pool temperature was defined (Section 4.3 of NUREG-0783) as "the relevant temperature which controls the behavior of the condensation process occurring at the quencher exit." In this context, the establishment of ΔT for the 12 blowdowns under consideration here will be described. It should be noted that these blowdowns are the most relevant for the present purpose because they provide the basis for determining the outermost boundary of the operating envelope.

Despite the fact that these blowdowns were performed in a small cell where very efficient mixing might be expected to occur, some evidence of temperature nonuniformity was exhibited by the data. For the five blowdowns for which this information was available, the maximum temperature differences indicated by the sensor array varied from a minimum of about 10°F to as much as 27°F. This variability makes it difficult to characterize the local pool temperature unambiguously, as was done by use of Sensor T11 in Reference 1. Adding to the difficulty is the fact that different values for the same temperature can be deduced from the several sources which are available. (See the values given in Table 6.7 of Reference 3, Figures 8.1 and 8.5 of Reference 1, and Figures 17 and 18 of Reference 2). Two specific examples illustrate these inconsistencies: for Test 223, the values 212°F, 214°F, and 217°F can be deduced as the maximums recorded by Sensor T11 from the three respective sources; in the same test, the peak temperature recorded by Sensor T56 can be found to be either 225°F or 216°F. Thus, the staff and its consultants at Brookhaven National Laboratory concluded that the most appropriate way to resolve these difficulties would be by spatial averaging.

In forming this average, the readings from Sensors T52, T53, T55 were excluded because they are directly between two quencher arms and thus are likely to sense nonconservatively high temperatures. The inclusion of the readings from the tank bottom was also considered to be a conservatism.

The average values obtained essentially eliminated the differences exhibited by the graphical data given in References 3 and 4. That is, for any given run the spatially averaged temperatures from the two sources differed by less than 2°F. Comparison of these averages with the tabular values given for Sensor T11 showed that the latter was on the average about 4°F higher. Thus, in those tests for which only the tabulated values of T11 were available, the local temperature was taken to be that value minus 4°F; this provided some additional conservatism. For the remaining tests, the average values, as defined above, were used as the local pool temperature.

References

*Nonproprietary version available for inspection and copying, for a fee, in the NRC Public Document Room, 1717 H Street, N.W., Washington, D.C. 20555.

1. KWU, "Condensation and Vent Clearing Tests in GKM With Perforated Pipes," KWU Technical Report KWU/E3-2594, May 1973*

2. KWU, "Construction and Design of the Relief System With Perforated Pipe Quencher," KWU Technical Report KWU E3/E2-2703, July 1971.*

3. U.S. Nuclear Regulatory Commission, "Mark II Containment Lead Plant Program Load Evaluation and Acceptance Criteria," USNRC Report NUREG-0487, October 1978, and Supplement 1, September 1980 (available from NRC/GPO Sales Program, Washington, D.C. 20555).

4. O. Voight and others, "Consequences Drawn From a Stuck Relief Valve Incident at Wurgassen Power Plant," presented at Second International Conference on Structural Mechanics in Reactor Technology, S ptember 10-14, 1973, in Berlin (available through public technical libraries).

NRC FORM 335 (7-77)	U.S. NUCLEAR REGULATORY COMMISSION **BIBLIOGRAPHIC DATA SHEET**	1. REPORT NUMBER (Assigned by DDC) NUREG-0783

4. TITLE AND SUBTITLE (Add Volume No., if appropriate) Suppression Pool Temperature Limits for BWR Containment	2. (Leave blank)
	3. RECIPIENT'S ACCESSION NO.

7. AUTHOR(S) T. M. Su and others	5. DATE REPORT COMPLETED	
	MONTH July	YEAR 1981

9. PERFORMING ORGANIZATION NAME AND MAILING ADDRESS (Include Zip Code) U. S. Nuclear Regulatory Commission Office of Nuclear Reactor Regulation Division of Safety Technology Washington, D. C. 20555	DATE REPORT ISSUED	
	MONTH November	YEAR 1981
	6. (Leave blank)	
	8. (Leave blank)	

12. SPONSORING ORGANIZATION NAME AND MAILING ADDRESS (Include Zip Code) U. S. Nuclear Regulatory Commission Office of Nuclear Reactor Regulation Division of Safety Technology Washington, D. C. 20555	10. PROJECT/TASK/WORK UNIT NO.
	11. CONTRACT NO.

13. TYPE OF REPORT Regulatory Report	PERIOD COVERED (Inclusive dates)

15. SUPPLEMENTARY NOTES	14. (Leave blank)

16. ABSTRACT (200 words or less) Boiling water reactor plants are equipped with safety/relief valves (SRVs) to protect the reactor from overpressurization. Operational transients, such as turbine trips, will actuate the SRV. Once the SRV opens, steam released from the reactor discharges through SRV lines to the suppression pool in the primary containment. Steam is condensed in the suppression pool in a stable condition. Extended steam blowdown into the pool, however, will heat the pool to a level where the condensation process may become unstable. This instability of steam condensation may cause severe vibratory loads on containment structures. Current practice restricts the allowable operating temperature envelope of the pool in the Technical Specifications so this instability will not occur. This restriction is referred to as the pool temperature limit. Task Action Plan A-39, "Determination of Safety-Relief Valve (SRV) Pool Dynamic Loads and Temperature Limits for BWR Containment," was established to resolve, among other things, the concern about steam condensation behavior for the Mark I, II, and III containments. This report presents the resolution of this issue including (1) acceptance criteria related to suppression pool temperature limits; (2) events for which a suppression pool temperature response analysis is required; (3) assumptions used for the analysis; (4) and requirements for the suppression pool temperature monitoring system. This report completes the subtask related to the suppression pool temperature limit in Task Action Plan A-39.

17. KEY WORDS AND DOCUMENT ANALYSIS	17a. DESCRIPTORS

17b. IDENTIFIERS/OPEN-ENDED TERMS Suppression Pool Temperature Limits

18. AVAILABILITY STATEMENT Unlimited	19. SECURITY CLASS (This report) Unclassified	21. NO. OF PAGES
	20. SECURITY CLASS (This page) Unclassified	22. PRICE S

UNITED STATES
NUCLEAR REGULATORY COMMISSION
WASHINGTON, DC 20555-0001

OFFICIAL BUSINESS

NUREG-0783

SUPPRESSION POOL TEMPERATURE LIMITS FOR BWR CONTAINMENTS

NOVEMBER 1981